LIFE BEGINS

On The Cover:
Ancient Ocean and Lightning.
Life began in the water—and stayed there for more than 3 billion years.
Cover Art by Walter Stuart.

Published by Creatvie Education, Inc., 123 South Broad Street, Mankato, Minnesota 56001

Printed by permission of Wildlife Education, Ltd.

ISBN 0-88682-387-0

Created and written by
John Bonnett Wexo

Chief Artist
Walter Stuart

Senior Art Consultant
Mark Hallett

Design Consultant
Eldon Paul Slick

Production Art Director
Maurene Mongan

Production Artists
Bob Meyer
Fiona King
Hildago Ruiz

Photo Staff
Renee C. Burch
Katharine Boskoff

Publisher
Kenneth Kitson

Associate Publisher
Ray W. Ehlers

LIFE BEGINS

This Volume is Dedicated to: Sigurd and Irene Wexo, my parents, who gave me life and much more than that. They created the possibility of all my books.

Art Credits

Pages Eight and Nine: John Francis; **Pages Ten and Eleven:** Timonthy Hayward; **Pages Twelve and Thirteen:** Timothy Hayward; **Pages Fourteen and Fifteen:** John Francis; **Pages Sixteen and Seventeen:** Walter Stuart; **Pages Eighteen and Nineteen:** Walter Stuart; **Pages Twenty and Twenty-one:** John Francis; **Pages Twenty-two and Twenty-three: Background,** Timothy Hayward; **Figures,** Chuck Byron.

Photographic Credits

Pages Six and Seven: Sigurg Jonasson.

Creative Education would like to thank Wildlife Education, Ltd., for granting them the rights to print and distribute this hardbound edition.

Contents

Something wonderful happened about $3\frac{1}{2}$ billion years ago . . .

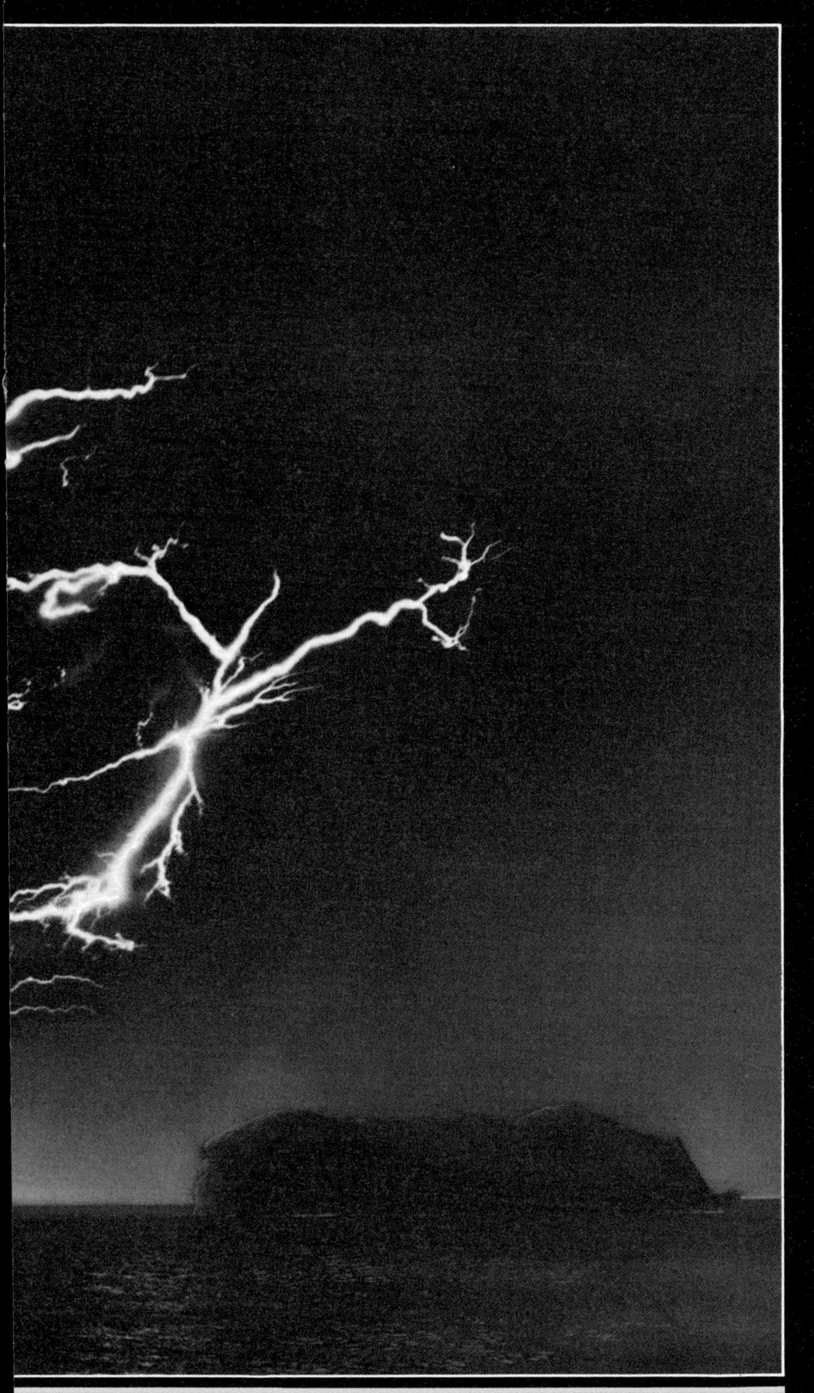

BOOK ONE

Life Begins

For a very very long time, there was **no life** on the Earth. There were oceans of water, and islands of rocky land, and huge storms to light up the sky—but not a single living thing. The Earth was formed about 4½ billion years ago, and it remained empty of life for almost one billion years after that. Then, about 3½ billion years ago, **the first living things** appeared. They were very small and simple plants, living in the water. And they were **very mysterious**. **Where** did they come from? **Why** did they appear? And **how** did they develop into the billions of living things that we see around us in the world today? These are some of the questions that scientists try to answer about life. As you will see, scientists have a unique way of trying to find the answers. . .

What is life? Exactly what is it that makes a living thing different from something that is not living? **What makes a rabbit different from a rock?**

This may seem like a simple question. After all, you live in a world that is full of living things—so it should be easy to decide what a living thing is.

Think for a moment. What do living things **do**? Most of them **move**, don't they? Is **this** what makes them different from non-living things? Or how about **growth**? All living things grow at some time in their lives. Is this the difference? Or how about **energy**? All living things take in some kind of food and use it to produce energy that runs their bodies.

As you will see, none of these can explain the difference between living and non-living things. The real difference may surprise you.

1ST WRONG ANSWER:

Some people think that only living things can move. They see birds and deer flying and running—and they are sure that non-living things cannot do these things.

2ND WRONG ANSWER:

Many people think that living things are the only things that can grow. They see small animals and humans get bigger —and they think that non-living things cannot do this.

In fact, many non-living things grow larger over time. Crystals are one fine example of this. By adding new minerals, tiny crystals can grow into huge crystals. But crystals are not alive.

3RD WRONG ANSWER:
Living things can turn chemicals from their food into energy. Some people are sure that only living things have this ability.
But waterfalls can run, and dead leaves can fly through the air. In fact, the air itself is moving all the time. Many things move that are not **alive**.
But the sun can turn chemicals into energy. In fact, the sun is the first source of **all** the energy that is used on earth—but the sun is not **alive**.
THE SURPRISING ANSWER:
The one big difference between living and nonliving things is **the ability to reproduce**. Living things can make **copies** of themselves—copies that look very similar and have similar abilities.
Living things can grow, and move, and make energy from food. But most important—**they can reproduce themselves.**

How did life begin? And why are there so many kinds of living things on earth? Scientists try to find answers to difficult questions like this by using a special way of finding facts that is called **the scientific method**.

First, they look for clues like detectives do. Then they study the clues and come up with **a possible reason** why something is the way it is. This is called **a hypothesis** (high-POTH-uh-sis).

Finally, they test the hypothesis to see if it is wrong. The tests are called **experiments**—and scientists may have to do many experiments to test a hypothesis.

EXPERIMENTS

Scientists may do **many different tests** to see if a hypothesis is wrong.

If many tests fail to show that a hypothesis is wrong, scientists can begin to assume that **it may be right.**

LIKE DETECTIVES

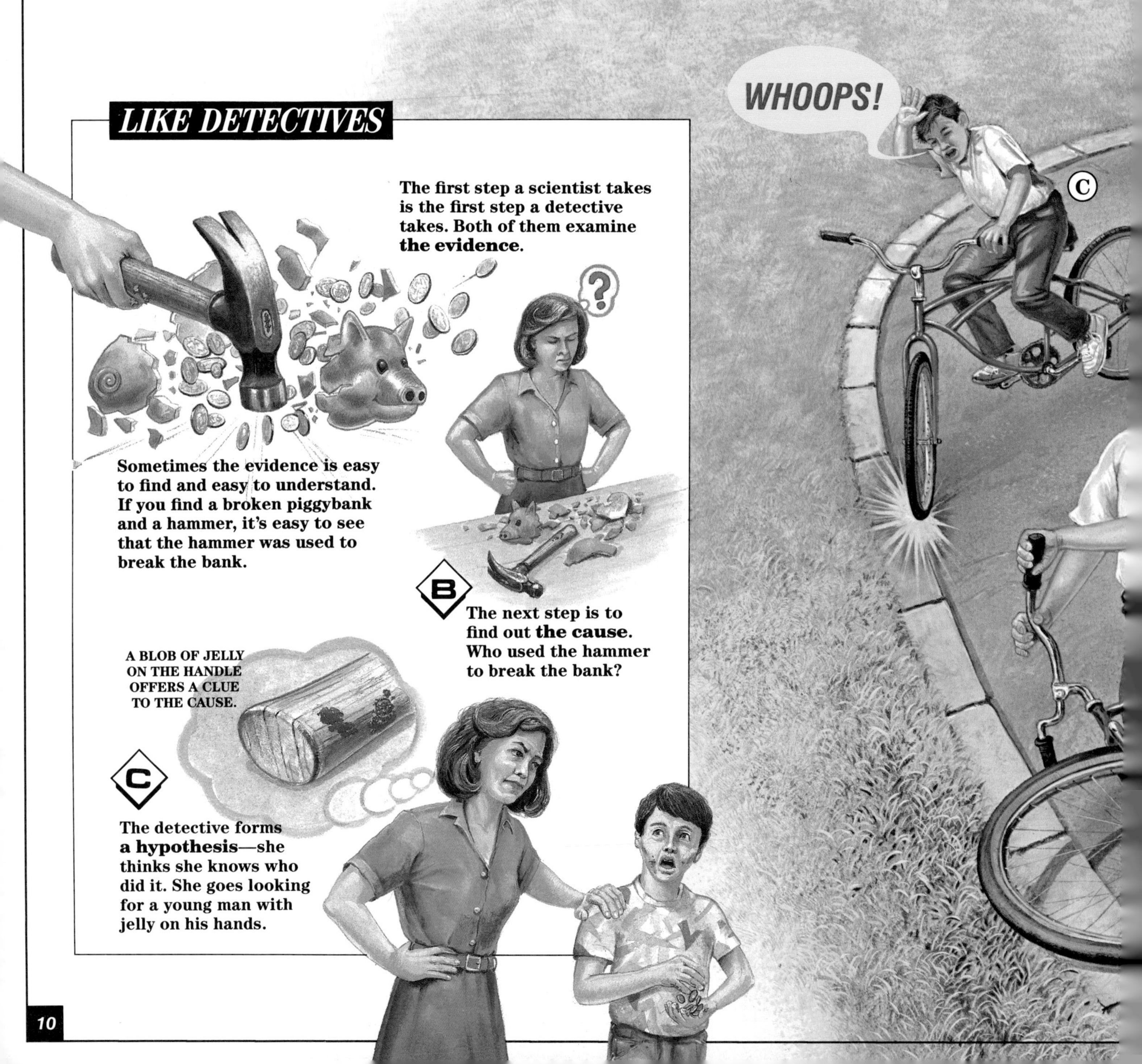
WHOOPS!
C
A BLOB OF JELLY ON THE HANDLE OFFERS A CLUE TO THE CAUSE.

The first step a scientist takes is the first step a detective takes. Both of them examine **the evidence.**

Sometimes the evidence is easy to find and easy to understand. If you find a broken piggybank and a hammer, it's easy to see that the hammer was used to break the bank.

B

The next step is to find out **the cause.** Who used the hammer to break the bank?

C

The detective forms **a hypothesis**—she thinks she knows who did it. She goes looking for a young man with jelly on his hands.

TRY AND TRY AGAIN

1

You do experiments all the time, although you may not be aware of it. For example, you have to experiment to **learn to ride a bicycle** . . .

Before you actually try to ride a bicycle, you look at other people doing it Ⓐ and **you form a hypothesis** about how it might be done. Then **you test your hypothesis.**

2

3

Your first hypothesis may be **just push the pedals** Ⓑ. This is wrong, so **the experiment fails**—you fall off the bike Ⓒ.

4

Your next hypothesis says **push the pedals and steer** Ⓓ.

5

This works, and **your experiment is a success** Ⓔ. You have found out something about bicycle riding!

Every time you ride the bicycle and don't fall off, you get more and more confidence that **your hypothesis is true.**

SCIENTIFIC THEORIES

After a hypothesis is tested many many times by experiments, it may be called **a fact.** Many such facts are put together to build **a theory.** Very slowly, scientists may build a theory that tries to answer questions like **"How did life begin on earth?"**

Using the scientific method, scientists have discovered many wonderful things about life and the world we live in. The most amazing discovery may be that **everything works according to rules**—and these rules of nature haven't changed for **billions of years**.

This means that we can find out many things about the past **by studying the way things work today**. By studying the way rocks are made today, we can find out how rocks were made in the past. By studying the way plants and animals work today, we can find out a lot about the way that ancient plants and animals worked.

Rocks from billions of years ago have **the same structure** as rocks that are being formed today—so we know they were **made in the same way** that rocks are made today. Watching a volcano as it makes new rocks today, we know how volcanoes did it in the past.

ANCIENT FOSSIL PLANT

LIVING PLANT

Fossils show that many plants that lived millions of years ago were **built like plants living today**. So we can use our knowledge about plants today to find out how plants worked in the past.

Millions of years ago, dinosaurs breathed air just the way you do today. In fact, the oxygen that you breathe today is **the same oxygen** that dinosaurs breathed!

Scientists tell us that living and non-living things have always been made of **the same kinds of chemicals**. The chemicals that formed animals and plants millions of years ago are the same chemicals that form animals and plants today.

From the largest creatures that ever lived to the smallest, everything on earth has always been built of **the same building materials.**

Millions of years ago, **the bodies of animals** worked in the same general way that the bodies of animals work today. For example, the skeletons of some dinosaurs were similar to the skeletons of dogs living today. By studying the way a dog's skeleton works, we can find out a lot about the way that the dinosaur skeletons probably worked.

Scientists have discovered that everything on earth is made of little pieces. These little pieces are called **atoms** (AT-ums)—and they are very very very small. They are so small that you cannot see them. In fact, **millions** of them can fit on the **point** of a pin!

Everything you see around you is made of millions and billions of atoms stuck together. Trees are made of atoms. Rocks are made of atoms. Dogs are made of atoms. This book is made of atoms, and even **you** are made of atoms.

ATOMS & MOLECULES

As small as atoms are, they are made of even smaller pieces called **particles** (PART-uh-culs). Some particles are in the center of an atom, and they form the **nucleus** (NEW-klee-us) of the atom.

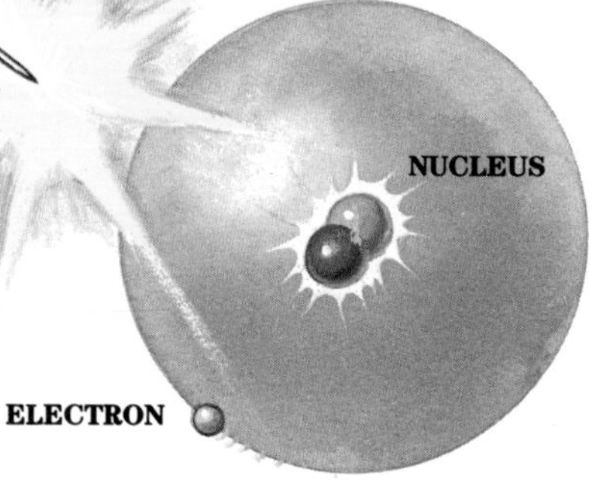

Other particles spin around the nucleus, like planets do around the sun. These particles are called **electrons** (ee-LEK-trons).

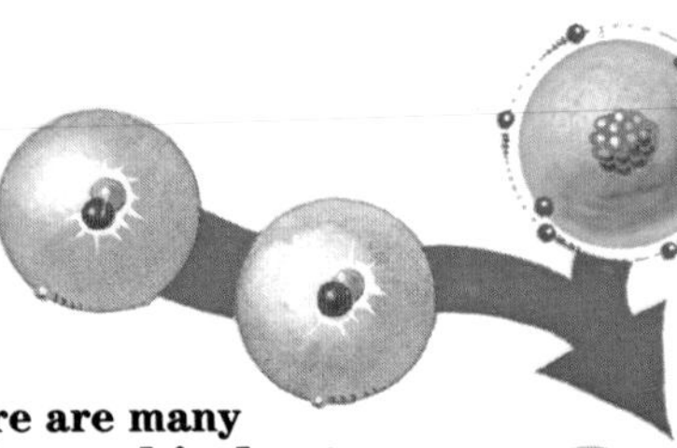

There are many **different kinds** of atoms. Each kind has a different number of particles in its nucleus, and a different number of electrons. These different kinds of atoms often **join together** to form larger groups of atoms called **molecules** (MOLL-uh-kyouls).

INVISIBLE FORCES

Everything on earth is held together by invisible forces—forces you cannot see. The gravity that holds you on the earth is one of these forces. You can't see it, but without it you would fall off the earth.

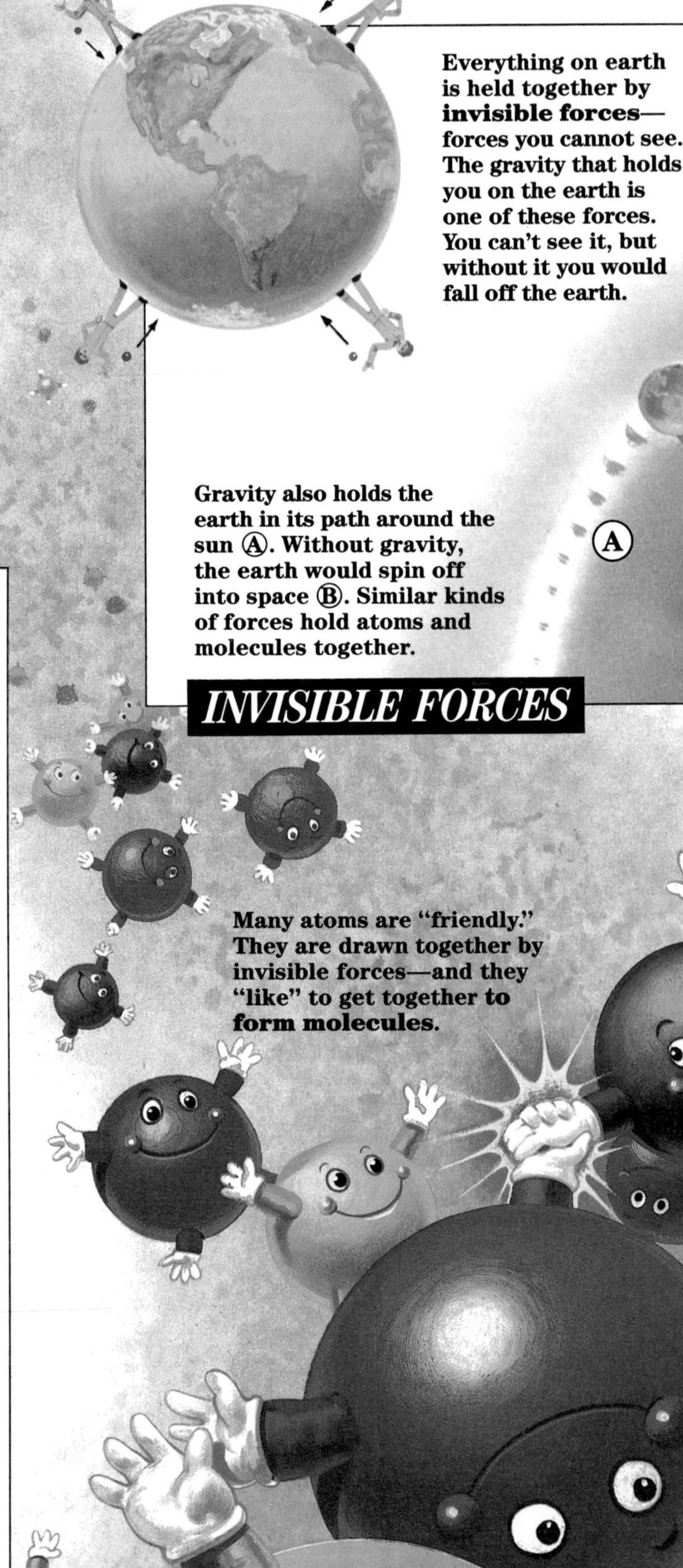

Gravity also holds the earth in its path around the sun Ⓐ. Without gravity, the earth would spin off into space Ⓑ. Similar kinds of forces hold atoms and molecules together.

Many atoms are "friendly." They are drawn together by invisible forces—and they "like" to get together to form molecules.

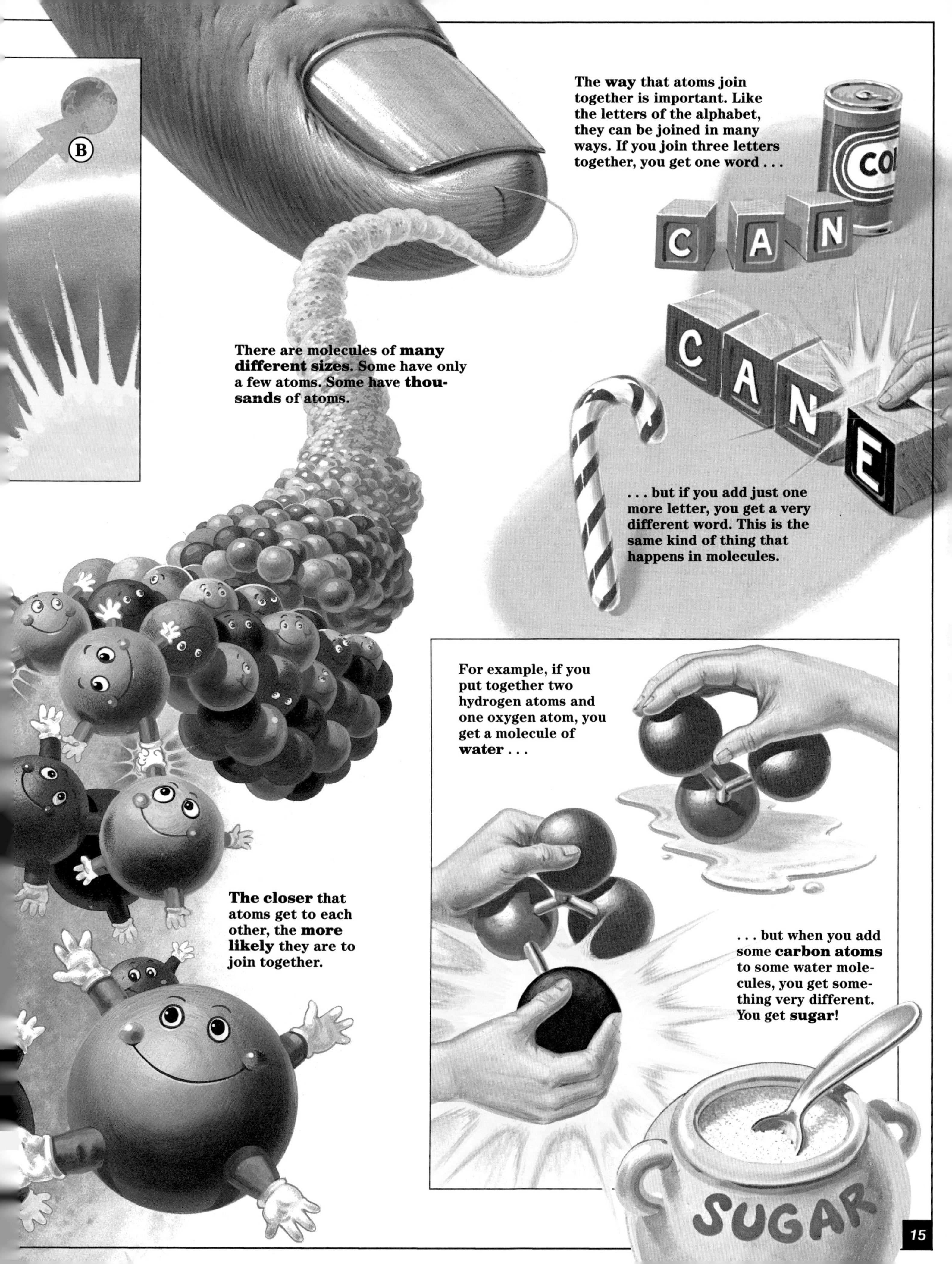

The **way** that atoms join together is important. Like the letters of the alphabet, they can be joined in many ways. If you join three letters together, you get one word . . .

. . . but if you add just one more letter, you get a very different word. This is the same kind of thing that happens in molecules.

There are molecules of **many different sizes**. Some have only a few atoms. Some have **thousands** of atoms.

The closer that atoms get to each other, the **more likely** they are to join together.

For example, if you put together two hydrogen atoms and one oxygen atom, you get a molecule of **water** . . .

. . . but when you add some **carbon atoms** to some water molecules, you get something very different. You get **sugar**!

The **whole universe** is made of atoms and molecules. Every star and every planet is made of the same kinds of building blocks that the earth is made of—and **you** are made of. In fact, most scientists believe that all of the atoms and molecules in the universe came **from one place**.

They believe that the universe began in **a big explosion** that scattered parts of atoms over a huge area in space. After a long time, the parts started **to join together** to form atoms, and then molecules. After a very long time, the atoms and molecules **formed stars and planets**.

SOLID, LIQUID & GAS

ICE IS SOLID

(A)

Everything is made of atoms and molecules. If the molecules are close together Ⓐ, they form **a solid**.

If the molecules are further apart Ⓑ, they can slip past each other. We call this "slippery" condition **a liquid**.

WATER IS LIQUID

WATER VAPOR IS GAS

When molecules get far away from each other Ⓒ, they become **a gas** that is very light like air.

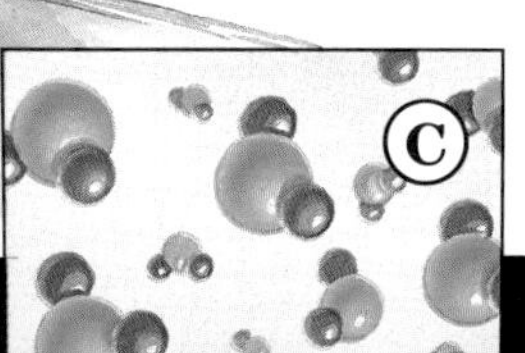

The clouds covered millions of miles of space. As time passed, the molecules in the clouds continued to get closer and closer together. After many millions of years, the molecules were close enough to form clouds of **dense gas** (shown below).

THE SUN

One of these dense clouds was **the beginning of our system**. In the middle of the cloud, the atoms in billions of molecules were torn apart and joined back together to form **the Sun**. A great deal of heat was produced as this happened. The sun **started to "burn."**

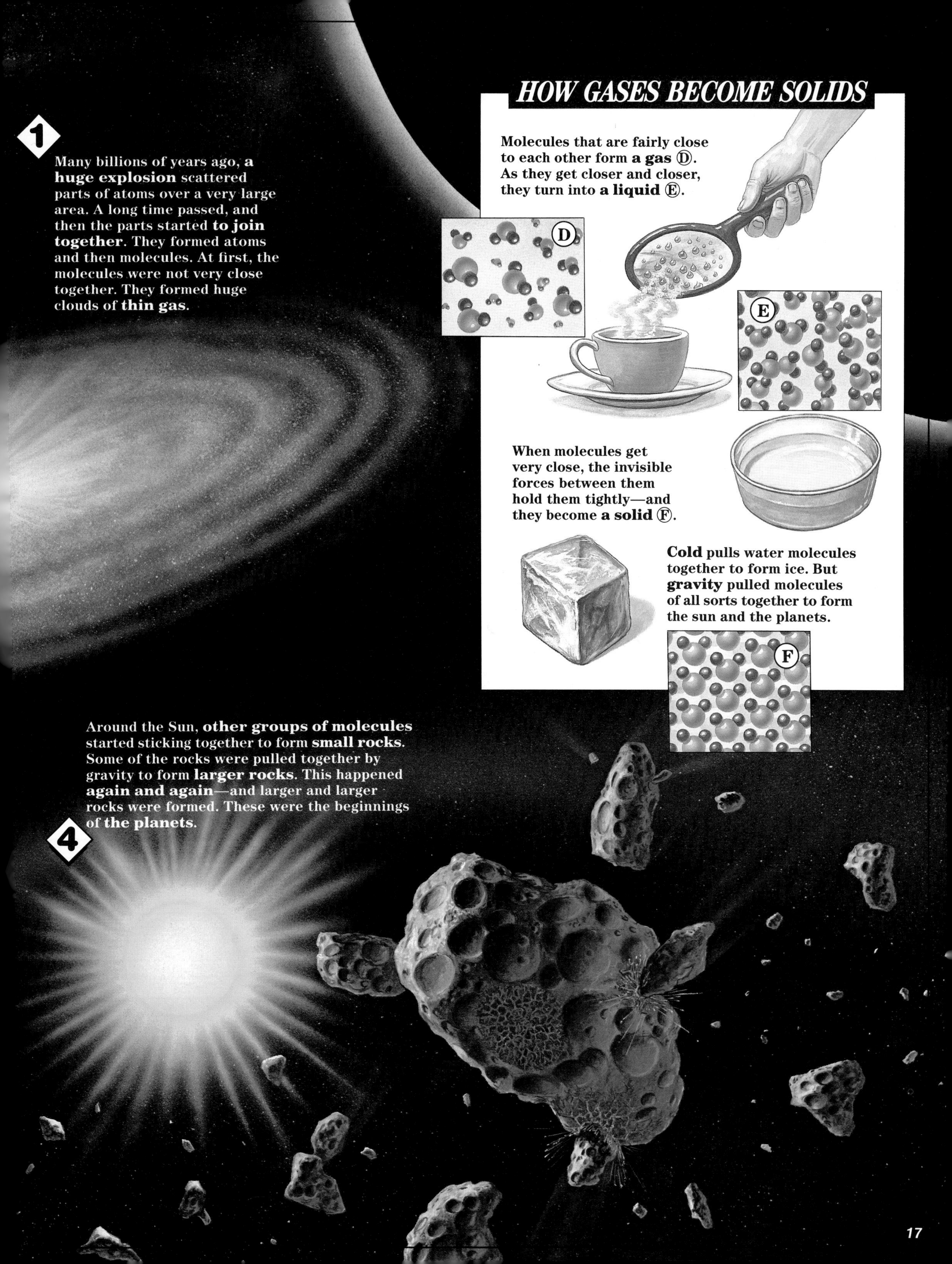

1

Many billions of years ago, **a huge explosion** scattered parts of atoms over a very large area. A long time passed, and then the parts started **to join together**. They formed atoms and then molecules. At first, the molecules were not very close together. They formed huge clouds of **thin gas**.

HOW GASES BECOME SOLIDS

Molecules that are fairly close to each other form **a gas** Ⓓ. As they get closer and closer, they turn into **a liquid** Ⓔ.

When molecules get very close, the invisible forces between them hold them tightly—and they become **a solid** Ⓕ.

Cold pulls water molecules together to form ice. But **gravity** pulled molecules of all sorts together to form the sun and the planets.

4

Around the Sun, **other groups of molecules** started sticking together to form **small rocks**. Some of the rocks were pulled together by gravity to form **larger rocks**. This happened **again and again**—and larger and larger rocks were formed. These were the beginnings of **the planets**.

The Earth was formed very slowly, over billions of years. Rocks were joined together with other rocks by the force of **gravity**. Smaller rocks with smaller gravity were pulled in by larger rocks with larger gravity. Finally, all of the rocks were joined into **one great big rock** that we call the planet Earth.

As you will see, the early Earth was not a very good place for life. For one thing, the surface of the planet was **boiling hot**. The rocks on the surface were so hot that they were liquid—like the lava that comes out of volcanoes.

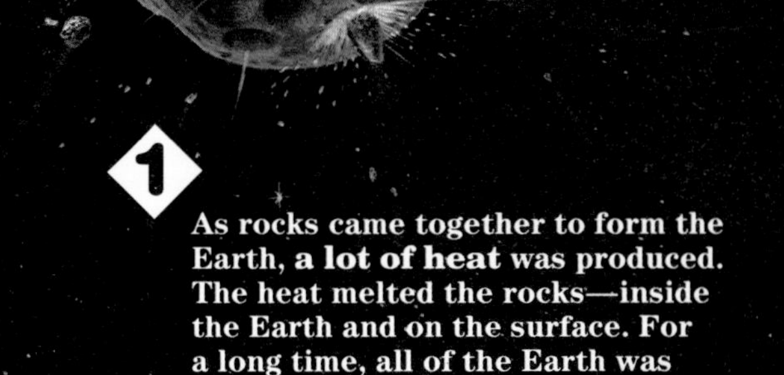

1

As rocks came together to form the Earth, **a lot of heat** was produced. The heat melted the rocks—inside the Earth and on the surface. For a long time, all of the Earth was **liquid rock**. It was **red hot**.

2

Inside the liquid rock, **atoms mixed together** in many different combinations. They mixed again and again and again. They formed many different kinds of **chemical molecules**.

THE FIRST ATMOSPHERE

As the hot rock came out of the volcanoes, it released **many kinds of chemical molecules.** These molecules became **the first atmosphere** of the Earth Ⓒ.

There were **many cracks** Ⓐ in the crust as the Earth started to cool. **Hot liquid rock** inside the Earth pushed up through the cracks and exploded out of thousands of volcanoes Ⓑ.

Many of the molecules in the atmosphere were **carbon dioxide** and **water vapor**.

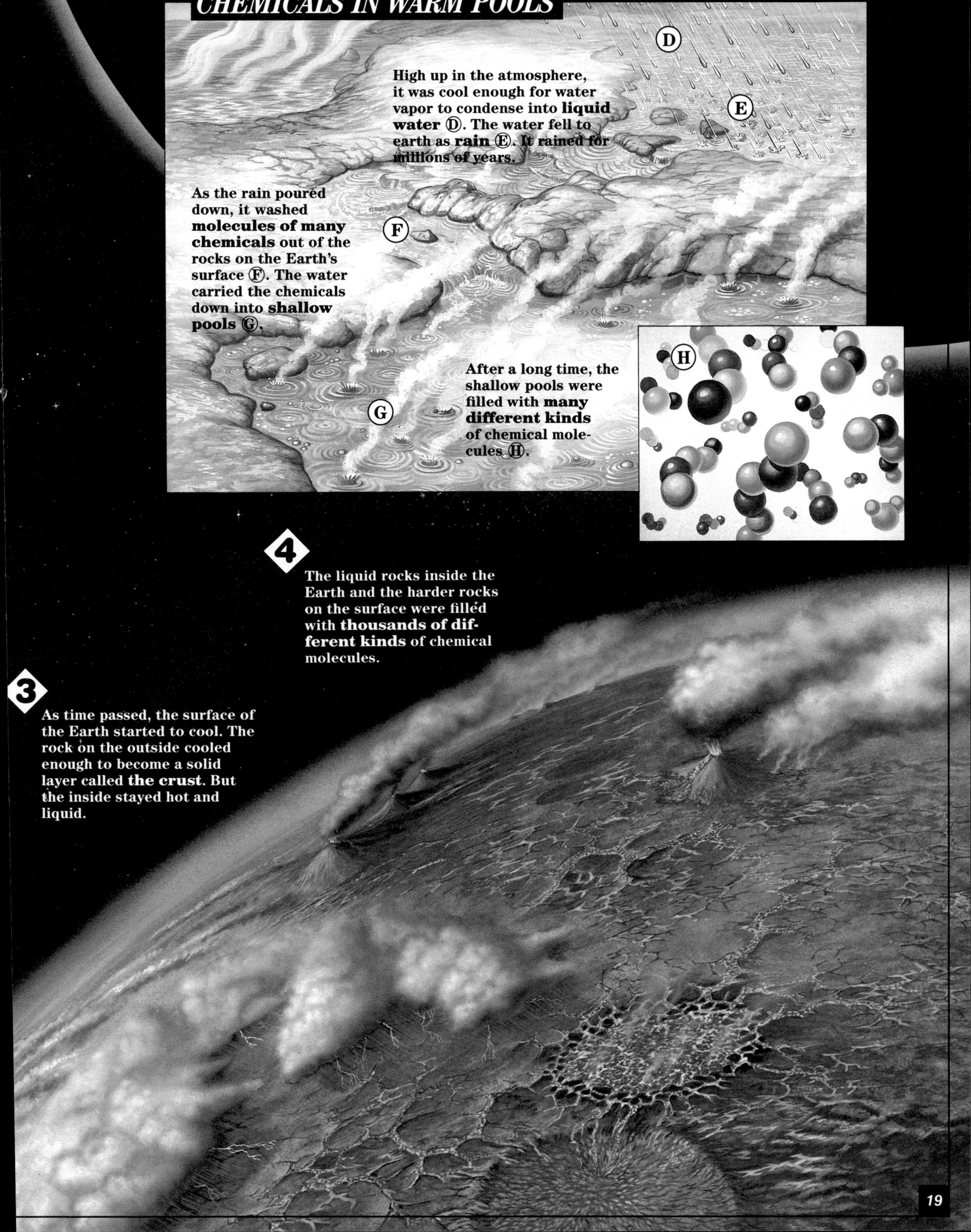
CHEMICALS IN WARM POOLS
D
E
High up in the atmosphere, it was cool enough for water vapor to condense into **liquid water** Ⓓ. The water fell to earth as **rain** Ⓔ. It rained for millions of years.
As the rain poured down, it washed **molecules of many chemicals** out of the rocks on the Earth's surface Ⓕ. The water carried the chemicals down into **shallow pools** Ⓖ.
F
G
H
After a long time, the shallow pools were filled with **many different kinds** of chemical molecules Ⓗ.
4
The liquid rocks inside the Earth and the harder rocks on the surface were filled with **thousands of different kinds** of chemical molecules.
3
As time passed, the surface of the Earth started to cool. The rock on the outside cooled enough to become a solid layer called **the crust**. But the inside stayed hot and liquid.

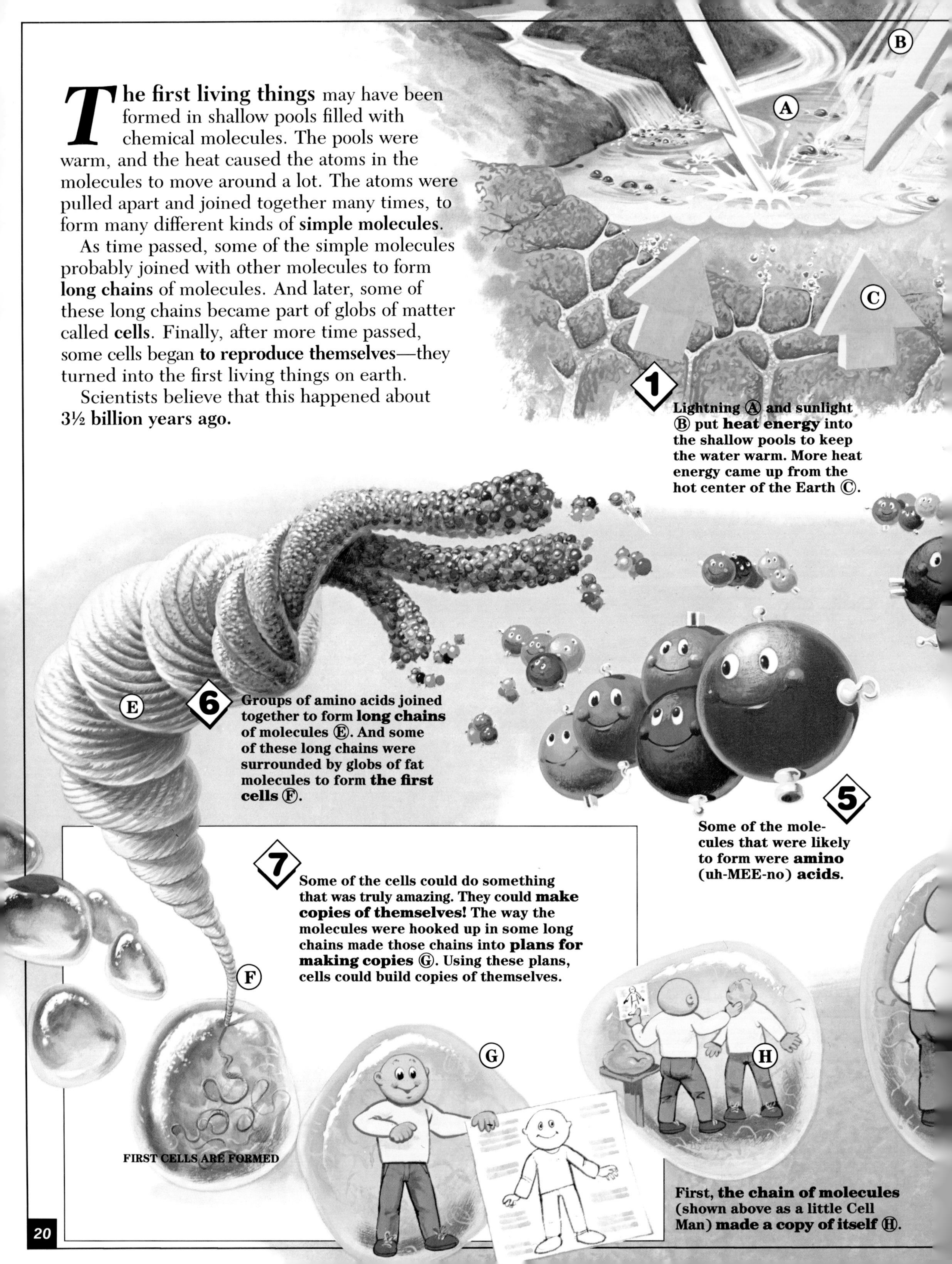

The first living things may have been formed in shallow pools filled with chemical molecules. The pools were warm, and the heat caused the atoms in the molecules to move around a lot. The atoms were pulled apart and joined together many times, to form many different kinds of **simple molecules**.

As time passed, some of the simple molecules probably joined with other molecules to form **long chains** of molecules. And later, some of these long chains became part of globs of matter called **cells**. Finally, after more time passed, some cells began **to reproduce themselves**—they turned into the first living things on earth.

Scientists believe that this happened about **3½ billion years ago.**

1 Lightning Ⓐ and sunlight Ⓑ put **heat energy** into the shallow pools to keep the water warm. More heat energy came up from the hot center of the Earth Ⓒ.

5 Some of the molecules that were likely to form were **amino** (uh-MEE-no) **acids**.

6 Groups of amino acids joined together to form **long chains** of molecules Ⓔ. And some of these long chains were surrounded by globs of fat molecules to form **the first cells** Ⓕ.

7 Some of the cells could do something that was truly amazing. They could **make copies of themselves!** The way the molecules were hooked up in some long chains made those chains into **plans for making copies** Ⓖ. Using these plans, cells could build copies of themselves.

First, **the chain of molecules** (shown above as a little Cell Man) **made a copy of itself** Ⓗ.

2

The heat energy forced the chemical molecules in the water to change again and again. The atoms in the molecules broke apart and formed **new kinds of molecules.**

3

Over and over again, molecules were broken up and new kinds of molecules were formed. Over millions of years, **many different combinations** of atoms were put together.

4

As you know, atoms like to join together to form molecules. When they get close enough, they "reach out" to grab each other. But atoms **are more "friendly"** with some atoms than with others—they "choose" their friends. They "prefer" to connect with some atoms and not with others Ⓓ. This means that some kinds of molecules were **more likely to form** than others in the warm pools.

The cell could **reproduce itself.**

You remember that living things are the only things on earth that can reproduce themselves—like rabbits. The cells could reproduce themselves. They were **alive.**

hen it made **another cell** Ⓘ. o hold the copy. The two cells plit apart—and there were **two dentical cells.**

REMEMBER:

1 Most of the time, you can easily tell something that is living from something that is not. But many people don't really know what the difference between them is.

2 Non-living things can sometimes grow, or move, or even turn chemicals into energy. But only living things can **make copies of themselves.**

3 To study life, scientists use **the scientific method**. They do **experiments** to find out how life might work.

4 First they form **a hypothesis**—and then **they test the hypothesis**. This is the same kind of thing that you do when you learn to ride a bicycle. If a test shows that one hypothesis is wrong, they form a new one and test that.

5 Using the scientific method, scientists have discovered that the world works **according to rules**. This means we can study the way things work today to find out how they worked millions of years ago.

6 Scientists have discovered that everything on earth is **made of atoms.**

7 Atoms are "friendly"—**they like to join together**. When atoms join, they form **molecules**.

8

The way that atoms j makes a difference. For ample, when you add ca bon atoms to water ato it can make **sugar**!

NEW WORDS:

Hypothesis

(high-POTH-uh-sis):
An attempt to explain **how something might work**. After scientists examine all the evidence they can find, they form a hypothesis.

Experiment

(ex-PEAR-uh-ment):
A test to see if a hypothesis is wrong.

Molecules

(MOLL-uh-kyouls):
Two or more atoms joined together.

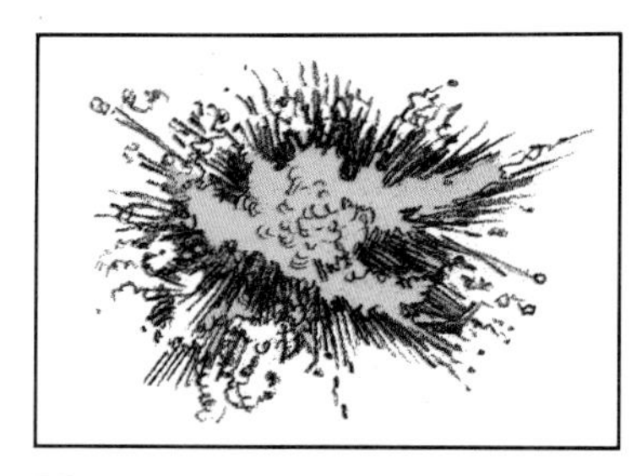

9 **The whole universe** is made of atoms and molecules. The universe began billions of years ago, with a big explosion that made huge **clouds of gas.**

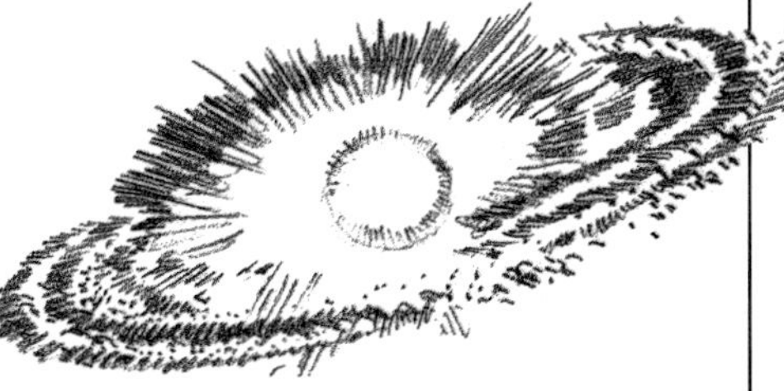

10 Over billions of years, the clouds of gas formed dense rings—and these became stars, **like the Sun.** Planets began to form around the Sun.

11 Gradually, dense gases turned to solid rocks—and the rocks joined together to form planets, like **the Earth.** When the Earth first formed, it was red hot.

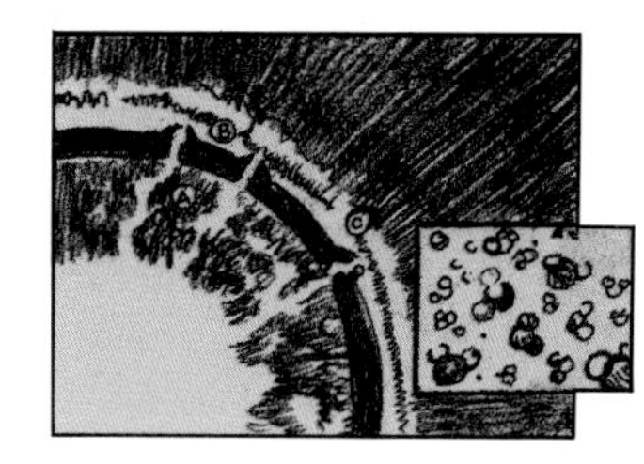

12 As the earth cooled on the outside, volcanoes brought hot gases full of chemicals up to the surface. These formed the first atmosphere. The atmosphere and the rocks on the surface were **full of chemicals.**

13 It began to rain, and it rained for **millions of years.** The rain washed chemicals out of the atmosphere and the surface rocks—and these chemicals were carried into **shallow pools.**

14 In the shallow pools, atoms were forming many different kinds of molecules. After a time, they formed special molecules called **amino acids.**

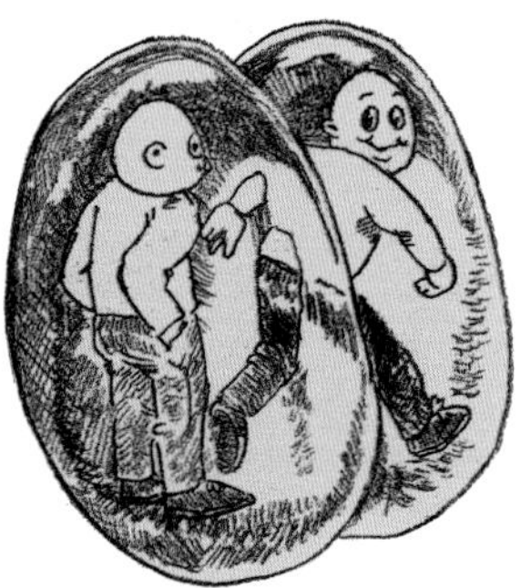

15 The amino acids became part of complex molecules that could **make copies of themselves**—and these molecules became part of **cells.** These cells could make copies of themselves.

16 The cells could do what rabbits and all other living things can do—the cells were **the very first living things!**

ms
ums):
very, very small pieces that
ything is made of. There are
lions of atoms on the point of a pin!

Particles
(PART-uh-culs):
The even smaller pieces that atoms are made of. The **nucleus** of an atom is made of particles.

Nucleus
(NEW-klee-us):
The group of particles that forms **the center** of an atom. The combination of particles in the nucleus of each kind of atom is different.

Index

Amino acids, 20
Ancient times, method of studying,12
Animal bodies, studying, 13
Atmosphere, formation of, 18
Atoms, 14
  formation of molecules from, 21
  joining together of, 14-15

Bones, studying, 13
Building materials of the universe, 16

Carbon atoms, 15
Carbon dioxide, in the Earth's atmosphere, 18
Cells
  formation of, 20
  reproduction of, 20-21
Chains of molecules, 20
Chemical molecules
  formation of, 18
  in shallow pools, 19
Chemicals, in living and non-living things, 13
Cracks in the Earth's surface, 18
Crust of the Earth, 18, 19
Crystals, growth of, 8

Dense gas, 16
Dinosaur skeletons, studying, 13

Earth
  cracks in the Earth's crust, 18
  formation of, 18-19
  before life began, 7
Electrons, 14
Energy
  in the formation of living things, 20-21
  as produced by living things, 8
  the sun as a source of, 9
Evidence, examination of, 10

Experiments, 10
  method of conducting, 10-11

Fact, determination of, 11
First living things, 20
Formation of the Earth, 7
Fossil plants, 12

Gases, 16
  how gases become solids, 17
Gravity, 14, 17
  role in Earth's formation, 18
Growth, as a characteristic of living things, 8

Heat energy, in the formation of living things, 20-21
Hypotheses,10
  testing, 11

Ice, formation of, 17
Invisible forces, 14

Life
  characteristics of, 8
  formation of, 20-21
  how it began, 10
Liquid rocks inside the Earth, 18, 19
Liquids, 16
Living things
  appearance of, 7
  studying ancient living things, 13
  variety in, 10

Molecules, 14
See also Chemical molecules
  simple, 20
  sizes of, 15
Motion, as a characteristic of living things, 8

Nature, rules of, 12

Non-living things, differences from living things, 8
Nucleus, 14

Particles, 14
Planets, formation of, 16-17
Plants
  appearance of, 7
  studying fossil plants, 12

Reproduction
  as a characteristic of living things, 9
  of first cells, 20
Rock
  formation of, 17
  liquid rock in the Earth's center, 18
  studying fossil rock, 12

Scientific method, 10
Scientific theories, 11
Scientists, questions asked by, 7
Simple molecules, 20
Skeletons, studying, 13
Solids, 16
  how gases become solids, 17
Stars, formation of, 16
Sun
  as an energy source, 9
  formation of, 16-17

Temperature, of ancient Earth, 18
Theory, 11
Thin gas, 17

Universe, structure of, 16

Volcanic rock, 12

Water molecules, 15
Water vapor, 16
  condensing into rain, 19
  in the Earth's atmosphere, 18